AMERICA DEBATES™

AMERICA DEBATES
GLOBAL WARMING:
CRISIS OR MYTH ?

Matthew Robinson

rosen publishing's
rosen
central®

New York

Published in 2008 by The Rosen Publishing Group, Inc.
29 East 21st Street, New York, NY 10010

Library of Congress Cataloging-in-Publication Data

Robinson, Matthew.
America debates global warming: crisis or myth? / Matthew Robinson.—
1st ed.
    p. cm.—(America debates)
Includes bibliographical references and index.
ISBN-13: 978-1-4042-1925-0 (hardcover)
ISBN-10: 1-4042-1925-0 (hardcover)
1. Global warming—Juvenile literature. I. Title.
QC981.8.G56R62 2008
363.738'74—dc22

                                                    2007010931

Manufactured in the United States of America

On the cover: Left: In Antarctica, a large chunk of a melting ice forma-
tion falls into the water. Right: Rush-hour traffic fills the freeway in
Los Angeles, California.

CONTENTS

Introduction

Are Earth's slowly rising temperatures the result of humans' impact upon our planet, or are they simply nature's way? This is the question being debated by scientists worldwide in the twenty-first century. Since the early 1990s, the issue of global warming has turned into one of the most hotly contested debates in the scientific community. It has many Americans questioning whether recent natural disasters like Hurricane Katrina could be side effects of global warming. In fact, the entire world is turning its attention to this very serious topic.

But what is at stake in the global warming debate? If global warming is indeed caused by humans, and temperatures continue to rise, then disasters like Hurricane Katrina might be a sign

This satellite image taken from space shows the immense size of Hurricane Katrina as it spread destruction across America's Gulf Coast.

of worse things to come. On the other hand, it is possible that global warming is not a crisis and that the planet's rising temperatures are just a natural part of Earth's climatic cycles. If that's the case, then all of the time and money spent trying to prevent global warming will be wasted, and all of the fear caused by global warming will be pointless.

The stakes are high in the debate on global warming. If one side is right, life as we know it on our beautiful planet could be forever changed for the worse. If the other side is right, billions of dollars that could go to better use will be wasted. While people on both sides may disagree on some points, most can agree that it is extremely important to make the right decision as quickly as possible.

Chapter 1

What Is Global Warming?

There are three main concepts behind the global warming debate that we must understand in order to form an educated opinion on the matter. These three components are the greenhouse effect, global warming, and climate change. An enhanced greenhouse effect (which we will discuss momentarily) leads to a state of global warming, which, over time, leads to climate change. All three concepts can be seen two ways: in the contexts of a natural environment and an anthropogenic environment. An anthropogenic environment is simply a fancy way of saying an environment that was caused by human interference and not by nature alone.

THE GREENHOUSE EFFECT

The greenhouse effect is one of the reasons that human beings are able to survive on Earth. If it weren't for the greenhouse effect, the atmosphere could never be warm enough for us to live. Our planet would, in fact, be in a permanent ice age. The greenhouse effect is a good thing—a very good thing. Here's how it works.

The sun is our main source for light and warmth here on Earth. It sends out a huge amount of solar radiation (energy) to our planet every day. Almost half of the heat and light that hits our planet bounces back into space; the rest of it is absorbed by Earth's atmosphere, water, and landmasses. This absorbed energy from the sun warms the surface of Earth as well as the atmosphere, which is the layer of gases surrounding the planet. By itself, however, this absorbed energy cannot keep our planet warm enough to sustain life as we know it. We also rely on the effects of atmospheric gases known as natural greenhouse gases to keep our world warm enough for us to survive.

The main natural greenhouse gases are carbon dioxide (CO_2), water vapor, methane, and ozone. These gases help keep some of the energy from the sun inside our atmosphere by bouncing it back toward Earth, not allowing it to escape into space. If you go to an actual greenhouse at a plant nursery or in someone's garden, you'll see a similar concept at work. The glass ceiling of a greenhouse allows the sunlight to come into the greenhouse yet also serves as barrier to keep heat from escaping. This allows people to keep plants in an optimal growing environment

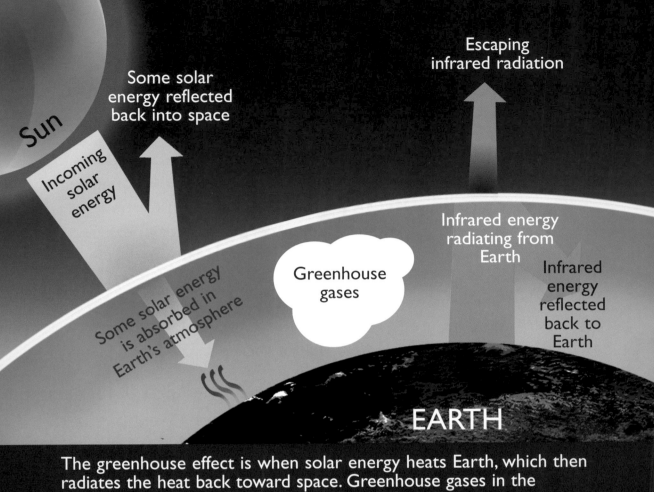

Sun

Incoming solar energy

Some solar energy reflected back into space

Escaping infrared radiation

Some solar energy is absorbed in Earth's atmosphere

Greenhouse gases

Infrared energy radiating from Earth

Infrared energy reflected back to Earth

EARTH

The greenhouse effect is when solar energy heats Earth, which then radiates the heat back toward space. Greenhouse gases in the atmosphere trap the heat, also known as infrared energy.

regardless of the climate outside. The natural greenhouse effect is a good thing for life on our planet.

A different type of effect is called the enhanced greenhouse effect. This occurs when the levels of greenhouse gases increase and trap more heat in Earth's atmosphere. Imagine if you added extra layers of glass to an actual greenhouse. The room would heat up quickly and probably create an environment harmful to the plants inside. An enhanced greenhouse effect may be caused by humans releasing large quantities of greenhouse

In 2005, more than one million people in southern China were affected by the region's most severe drought in fifty years. Global warming may lead to such extreme weather.

gases into the atmosphere. The main culprit that is thickening up our atmosphere is carbon dioxide, which is released from the burning of fossil fuels, such as gasoline and coal. When increased greenhouse gases trap more heat in our atmosphere, we get global warming.

GLOBAL WARMING

Global warming isn't much more complicated than it sounds. It is simply the slow warming of our planet's atmosphere and surface over time. Usually when you hear the phrase "global warming" on the news or at school, you can bet they are talking about global warming caused by human beings (anthropogenic). Global warming takes place naturally, too, but many scientists believe the type of global warming occurring today is very different from warming caused by nature. Those who think global warming is not a problem believe the warming taking place today is mostly part of a natural cycle.

Global warming leads to an increase in temperatures worldwide, which can lead to changes in weather patterns, increased or reduced sea levels, varying lengths of seasons, and strange, unpredictable weather in general. This is climate change.

CLIMATE CHANGE

Climate change, like global warming, isn't much more complicated than it sounds. It simply means a slow change in Earth's climate. Climates vary all over the planet. If you live in sunny Southern California, then you're used to mild winters, warm summers, and little rainfall. This is the climate in Southern California. If you live in Alaska, you're used to long, cold winters, tons of snow and rain, and cool, crisp summers. This is the climate in Alaska. If over a long period, the winters in Alaska began to get warmer and the summers in Southern California began to get cooler, then most likely the world would be undergoing climate change.

If the climate is changing, then it would be nothing new for our planet. Earth has been subjected to severe climate changes in

Climate describes the general weather conditions of an area over a long period. Great changes in the global climate on planet Earth (*opposite page*) take place over thousands of years.

the past. The many ice ages the planet has gone through were all severe climate changes. The times when the planet was coming out of an ice age and warming up were also severe climate changes. In fact, Earth is always going through some sort of climate change, spread out over thousands or even millions of years.

And yet, the type of climate change being experienced today has a lot of scientists, world leaders, and average citizens very worried. They believe there is a lot of reliable, scientific data that says the present climate change is different from any that has taken place before. Those on the other side of the argument see the data differently. They don't think Earth's present changes are any different, or more worrisome, than the climate changes that have been going on for millions of years.

We've now covered the three major concepts involved in the global warming debate and how each of them can be understood from either the "human-made" or the "natural" point of view. Since all three of these concepts could possibly be caused by humans or caused by nature, how are we supposed to make an educated decision on the global warming debate?

The best you can do is to understand the main pros and cons of the debate. In the following chapters, we'll take a look at our planet today and hear different opinions on whether the planet is actually in a state of dangerous global warming. We'll look at the different viewpoints regarding whether mankind could be to blame for this global warming. From there, we'll look at the different theories on what we can expect to happen to our world if human-made global warming is indeed taking place. Finally, we'll take a look at what, if anything, can be done to prevent global warming—or if anything needs to be done at all.

Chapter 2

Is Global Warming Happening Now?

J ust about every respectable scientist agrees that, yes, over the past twenty years, global temperatures have been rising. What some disagree on is whether this warming trend will continue, whether it is caused by human activity, and if it is anything to be worried about. The global warming debate is not a black-and-white issue. There are extreme positions on both sides, and there are more moderate positions, too. In addition, many smart, responsible people don't have any opinion at all on the subject. They think the best information on both sides is still inconclusive. Most observers, however, fall into either the "global warming is a crisis" camp or the "global warming is a myth" camp.

IT'S A CRISIS: GLOBAL WARMING IS TAKING PLACE RIGHT NOW . . . AND IT IS SERIOUS

Global temperatures have been precisely documented since 1860. Widely varying global temperatures have been recorded since that year. But did you know that out of the twenty-one hottest years (average global temperature) since 1860, twenty of them have taken place in just the past twenty-five years? And 2005 was the hottest of them all. This means that in the past 140 or so years, the last twenty-five have been unusually hot. But let's not forget that Earth is billions of years old. So, how much importance should we place on just the last 140 years?

Reading Ice Cores

Recently, scientists discovered a new way of using ice samples to read the planet's temperatures far earlier than just 1860. The most useful ice is found in Antarctica, Greenland, and the North and South poles, where it hasn't melted in hundreds of thousands of years. The scientists drill very deep to find ice that froze a long time ago. Within these ice samples, which they call ice cores, they are able to find air bubbles that were frozen in the snow and ice from up to 700,000 years ago. These bubbles tell them exactly how much CO_2 was in the air at that time, and they even indicate what the temperature was back then.

To many, the results of testing these ice cores have been very alarming. They show a world where temperatures have risen in the past thirty years like they've never risen before. Even looking at just the past thousand years, we can see temperatures have

This is what a typical ice core sample looks like after being drilled from a glacier. Scientists wear protective clothing and use sterilized tools to avoid contaminating the sample.

risen and fallen slightly, but never have they risen as extremely or as quickly as they have in the past thirty years.

Sea Levels

Another way to find out if global temperatures are rising is to look at sea levels. Sea levels refer to the average water levels in the oceans around the world. If you go to the beach and watch the tide come in, then you can understand sea levels. At high tide, the sea comes way up on the beach; at low tide, the ocean recedes. If over time you observed high tide coming up higher and higher, then you could say the sea levels were rising. You'd be right to assume that global temperatures were rising as well.

Rising sea level in coastal Florida

Areas shown in darkest tone could flood if global warming causes sea level to rise 2 ft. (0.6 m) in the next 100 years.

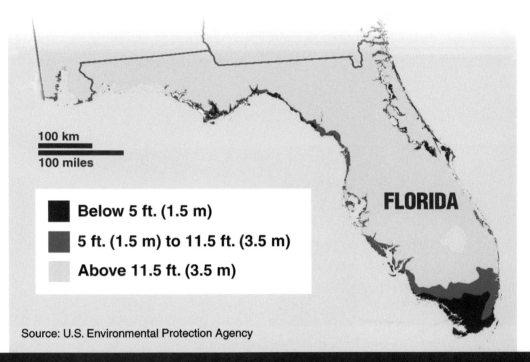

100 km

100 miles

■ Below 5 ft. (1.5 m)

■ 5 ft. (1.5 m) to 11.5 ft. (3.5 m)

□ Above 11.5 ft. (3.5 m)

FLORIDA

Source: U.S. Environmental Protection Agency

Rising sea levels affect coastal regions around the world. Much of Florida's coastline would lie under water if global warming caused sea levels to rise significantly.

What do higher sea levels have to do with rising global temperatures? Well, they are related in two ways. The first is a basic scientific principle: warm liquids expand. If you've ever seen a mercury thermometer rising in response to heat, or put your hot chocolate in the microwave only to take it out and find it overflowing, then you've seen this take place with your own eyes. On a giant global scale, it works just the same. If global temperatures rise, then the oceans will expand and the sea levels will rise. Scientists all over the world have recently noticed that sea levels, indeed, are rising.

The melting of sea ice and glaciers is another reason why global warming is related to rising sea levels. Most of the world's ice can be found in Antarctica, and a lot of that ice is floating on top of the water. What happens when that ice melts? It no longer floats on top of the ocean and instead flows into the ocean, thus increasing the amount of water and causing sea levels to rise. A lot of people are concerned about the melting of glacier ice. If a large sheet of it were to melt and fall off into the ocean, it could actually cause a huge wave called a tsunami that could potentially hurt a lot of people living along the coast.

In the air, in the ice, and in the water, there are many ways to tell that global temperatures are on the rise.

IT'S A MYTH: GLOBAL WARMING IS NORMAL AND NOTHING TO WORRY ABOUT

It's hard to argue against the fact that global temperatures are on the rise right now. But does that mean they'll continue to rise? Those who believe global warming is not a crisis think that the rising global temperatures are no big deal at all and that soon we might even start to see temperatures decrease again.

Even though they agree that global temperatures are rising, not all scientists agree on just how high these temperatures are rising. There is no consensus, either, on whether the amount temperatures are rising is even unusual. In fact, many scientists who are not alarmed by recent global warming think we should relax and enjoy the warm weather we're experiencing. After all,

Aerosol spray cans were thought to be a reason behind the "global cooling" many scientists believed was taking place in the 1970s.

who would complain about sunnier beach days, shorter winters, and longer summers?

We've Seen It All Before

Some skeptical scientists recall the environmental "crisis" that passed a mere thirty years ago, in the mid-1970s. At that time, many scientists believed we were experiencing a "global cooling" caused by aerosols, tiny particles suspended in the atmosphere. Natural processes like volcanoes and dust storms produce most of these particles. However, human activity, too, was thought to be responsible for producing some of these harmful aerosols. Smoke from burning fossil fuels and using aerosol spray cans, for example, were supposed to contribute to the cooling. By the late 1970s, most theories about the slow cooling of the planet had been dismissed. How big of a deal is today's global warming, when only thirty years ago we were worried about global cooling?

It's Not So Hot

Another argument used to reduce the fear of a global warming crisis focuses around a time known as the medieval warming period. From about 900 CE to 1100 CE, our planet experienced a rise in temperatures that brought the average global temperature up to 60–61 degrees Fahrenheit (16 degrees Celsius). For comparison, our average global temperature today is about 59°F (15°C). That means that about 1,000 years ago, our planet was actually hotter than it is now. Some scientists believe that for the past 1,000 years, our planet has been cooling off from this medieval warming period. The slight warming we're experiencing now, they say, is simply Earth warming up a little bit after a long cooling trend.

While both sides agree we are experiencing a warming period right now, you can see that both sides have very different ways of interpreting this information. Now let's take a look at an important part of the debate: Are humans responsible for global warming or not? In the following chapter we'll take a look at some of the main arguments from both sides and do our best to get to the bottom of this heated debate on global warming.

Chapter 3

Can Human Beings Cause Global Warming?

Earth is a truly awe-inspiring planet. Its surface area measures roughly 200 million square miles (518 million square kilometers); in tons, it weighs a number that starts with six and has twenty-one zeros after it (a million has only six zeros in it!); and it is estimated to be somewhere between three and five billion years old. To put it simply, Earth is very big, very heavy, and very old.

Modern humans, on the other hand, stand somewhere between five and six feet (1.5 to 1.8 meters) in height, average about 100 to 200 pounds (45 to 91 kilograms) in weight, and have been alive on planet Earth for only a few hundred thousand years. Compared to Earth, humans are small, light, and young.

So is it realistic to think that humans could actually affect something as massive as Earth in any significant way? And if so, could the past few hundred years really be long enough to permanently affect the planet? Earth has been around for billions of years, so how could humankind, in such a short time, possibly destroy the planet's ability to sustain life? All of this may sound like an exaggeration, but there are a lot of people out there—scientists, politicians, astronomers, and regular people like you and me—who think humans can radically alter our planet. Moreover, they believe we might actually be able to do it within a very short time.

Let's take a look at the two sides of this argument: those who think it's impossible for humans to have harmed Earth in so short a time, and those who think humans are indeed changing Earth's ecological balance.

IT'S A MYTH: HUMANS ARE NOT TO BLAME

Many who think dangerous global warming is a myth believe Earth's climate is a very complicated thing, one that we don't completely understand yet. We don't fully understand the cycle of ice ages that have come and gone during our planet's existence. We don't fully understand why CO_2 levels rise and drop over time, and we don't fully understand why temperatures rise and drop either. Those in the "myth camp" think the "crisis side" is simply overreacting to events taking place on our planet over a very short time. They say that if we look at the big picture, meaning Earth's entire existence, what is taking place right now simply is not unusual.

Not all of the CO_2 in our atmosphere comes from human activity. Volcanoes and other natural sources have been emitting CO_2 into the atmosphere for a long time.

CO_2 Levels Are Not a Good Indicator

The myth side points out that in the past, CO_2 levels in our atmosphere were higher than they are today, and therefore what's taking place isn't that unusual. Ice core samples show that hundreds of thousands of years ago, the CO_2 concentrations in the atmosphere were much higher than they are now, and humans weren't around back then to cause it. Perhaps this means that the planet has a natural CO_2 cycle that we don't fully understand—a CO_2 cycle that goes up and down completely independently of human actions.

In addition, skeptical scientists point to statistics that show human activity is responsible for only about 3 percent of the

Events taking place on the surface of the sun can affect Earth's climate. The sunspot seen here is actually six times wider than Earth itself.

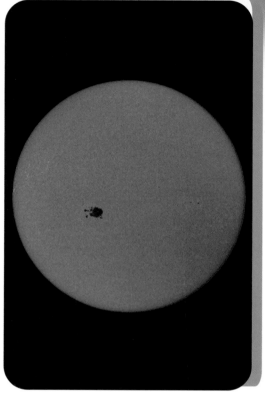

186 billion tons of CO_2 that enter the atmosphere annually. Such a small number cannot be the cause of the "crisis."

Sun Cycles

There's another factor that one must consider when thinking about rising temperatures on our planet, and it's an obvious one: the sun. The source of our planet's heat and light does not emit (give out) the same amount of heat and light all the time. Just as our planet has seasonal changes, the sun goes through changes as well. These changes include the coming and going of sunspots on the surface of the sun. Sunspots are sections of the sun's surface that are cooler than the rest. They look like dark splotches on the sun. While they can't be seen with the naked eye (don't try—you'll damage your eyes staring at the sun!), they can be seen with special telescopes.

Sunspots appear and disappear in a regular pattern known as the Schwabe solar cycle. The timing of the cycle indicates that the more sunspots there are, the hotter it is on Earth. That sounds backward, but it's true. At times when there are increased cold spots on the sun's surface, the sun's radiation hitting our planet is more intense. The cycle lasts an average of eleven years, and as of 2007, we were near the lowest point of the cycle, known as the solar minimum.

In 2000, we were at the highest point in the cycle, when the temperatures would be highest coming from the sun. This was exactly when a lot of people around the world began to worry about global warming and when a lot of scientists began to study it. What if the rising temperatures that some think are caused by CO_2 levels in the atmosphere are really just the sun heating up because of its sunspots? What if we're not looking at the big picture? That is the question most often asked by those on the myth side of the global warming debate.

IT'S A CRISIS: HUMANKIND IS TO BLAME FOR GLOBAL WARMING

The majority of scientists in the world today believe there is a connection between temperatures rising all over the world and humans' burning of fossil fuels. What are fossil fuels, and how do they raise global temperatures?

Fossil Fuels and Energy Production

Fossil fuels are what we use to power most of the technology that runs our world. The gasoline in our cars is a fossil fuel. Coal, oil, and natural gases are fossil fuels. Fossil fuels are in

Many industries power their factories by burning coal, a fossil fuel. Carbon dioxide (CO_2) and other greenhouse gases are then released into the atmosphere through smokestacks.

the fuel that propels a jet plane across the Atlantic Ocean. These energy sources are used to produce electricity for our houses, giving us light, heat, and the ability to run electronic devices like televisions and computers. Almost all modern technology relies on fossil fuels. Whether a given piece of technology uses fossil fuels as its energy source, or the machines that built that piece of technology were powered by fossil fuels, nearly everything in our world today is connected to fossil fuels.

So what's so bad about fossil fuels? Well, when you burn them to extract their energy, they release the natural gas carbon dioxide (CO_2). As we learned in chapter 1, CO_2 is one of the main natural gases floating in our atmosphere that make it possible for the greenhouse effect to take place. Think of CO_2 as one of the main layers of protection that keep enough heat inside our atmosphere to keep us all warm. Those who are concerned about global warming believe that it is harmful to burn too many fossil fuels over a long period of time. With too much CO_2 in the atmosphere, less heat escapes into space, and our Earth begins to warm up, perhaps to dangerous levels.

CO_2 in the Atmosphere

As we saw in the previous chapter, scientists are able to use ice core samples to find out how much CO_2 was in our atmosphere thousands of years ago. Using this and other methods, they are able to compare ancient levels of CO_2 to the amounts in our atmosphere today. Here's what they've found: For the last 10,000 years, the level of CO_2 in our atmosphere has been very consistent at around 280 parts per million. That sounds complicated, but all you really need to understand is that the number 280 has been pretty steady for 10,000 years.

Then, starting about 150 years ago, humans began burning fossil fuels in great quantities. The cause of this rapid increase was the Industrial Revolution. Put simply, the Industrial Revolution was a period beginning in the 1800s when large machines (like those found in factories) began to do the work previously done by humans. These machines ran on coal and other fossil fuels, so humans began burning a lot more fossil fuels—and dumping a lot more CO_2 into the atmosphere.

Since the Industrial Revolution, the amount of CO_2 in our atmosphere has jumped from 280 parts per million (ppm) to 380 ppm. This is the highest concentration of CO_2 in our atmosphere in more than 400,000 years. In about 150 years of heavy industry, the CO_2 levels in our atmosphere have risen higher than they've been in 400,000 years! And if things continue

THE INTERGOVERNMENTAL PANEL ON CLIMATE CHANGE

The Intergovernmental Panel on Climate Change (IPCC) was created in 1988 by the United Nations and the World Meteorological Organization. Its purpose is to study climate change and determine whether humans cause global warming and, if so, what that could mean for our planet.

Made up of more than 2,500 scientists, the IPCC issues regular reports on the state of climate change in the world. Its last full report (2001) stated that there is strong evidence that a lot of the warming seen all over our planet is indeed caused by greenhouse gases pumped into our atmosphere through human activity.

Certain sections of the IPCC's 2007 report have already been published, and they show global temperatures continuing to rise. The latest report offers even more proof that the majority of the rising temperatures have taken place in the last fifty years and are caused by humans' use of fossil fuels and the production of greenhouse gases.

to stay the same as they are now, that number is projected to rise as high as 560 ppm by the middle of the century.

A CO_2 concentration of 560 ppm could cause the temperatures on our planet to rise anywhere from 3 to 7°F (1.7 to 3.9°C). While that might not sound like such a big deal, consider this: the global temperatures that caused the last ice age were about 5 to 9 degrees colder than they are now. If a 5 to 9°F (2.8 to 5°C) shift in temperature in one direction can cause a global ice age, what effect do you think a shift of 3 to 7 degrees in the other direction could do? Here's one thing we know: it probably won't be all good.

Many scientists today are nearly certain that global warming is caused by humans and that the situation is indeed getting worse. Among those who believe this are those serving on the IPCC (Intergovernmental Panel on Climate Change), the world's largest collection of scientists dedicated to understanding global warming and climate change.

Deforestation and Global Warming

Outside of the burning of fossil fuels, there are a few other human-produced problems taking place. One of them is something called deforestation. This is the destruction of large expanses of woodlands, jungles, and rain forests. These natural habitats are often destroyed to make more room for livestock or farming, or to build areas for people to live and work in. As the population of our planet grows, the amount of large woodlands, jungles, and rain forests shrinks. But how does this affect global warming?

As we know, trees produce oxygen. If it weren't for trees, there wouldn't be any air to breathe, and human beings wouldn't be able to survive on our planet. But trees also serve another important purpose: they soak up some of the carbon dioxide in our atmosphere. This is done in a process called photosynthesis. Basically, the trees take in carbon dioxide, store up carbon in their plant tissues, and release oxygen for us to breathe.

Every year, millions of trees in Brazil's Amazon rain forest are lost to deforestation. These trees remove carbon dioxide from the atmosphere and produce the oxygen we breathe.

With fewer trees on our planet due to deforestation, there is more carbon dioxide in our atmosphere. And since the amount of CO_2 we're pumping into the atmosphere is already potentially dangerous, we're making the problem worse by cutting down the trees that soak up the important greenhouse gas. It's like a one-two punch to the environment: more CO_2 in the atmosphere and fewer trees to help soak it up. According to those on the crisis side of the argument, the current state of affairs could eventually make our planet a dangerous place to live.

Now that we've heard from both sides on whether humans are to blame for global warming, it's time to take a look at what is really at stake in this argument. If those on the crisis side are right, what does it mean for us and our world? How bad could things get? And if those on the myth side are correct, why are we worrying over nothing? Why are we wasting time, money, and resources trying to fix a problem that doesn't exist?

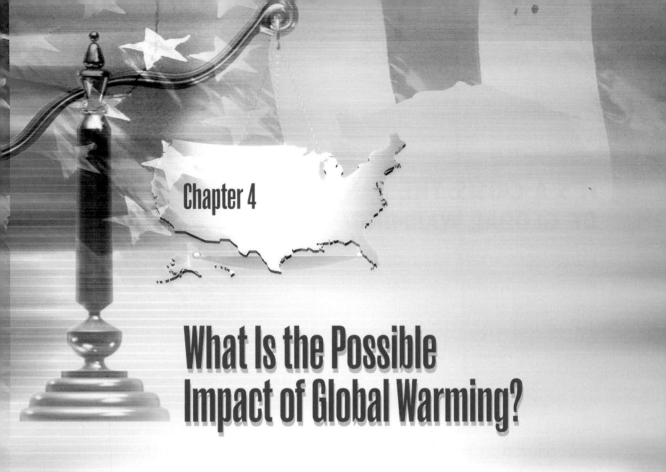

Chapter 4

What Is the Possible Impact of Global Warming?

Nobody likes to think about scary things that may or may not take place in the future. If global warming is caused by humans, and if it continues to get worse, the crisis side of the argument tells us that our lives could be greatly affected in the not-too-distant future. All of us would much rather believe that global warming is not caused by human activity, that it's natural, and that soon our weather will get back to normal. In fact, many on the "myth" side of the debate argue that we are better off working on problems that affect our world today, instead of worrying about things that may or may not affect us in the future. But whether we want to believe it or not, it's important to understand the possible outcomes of global warming.

IT'S A CRISIS: THE DANGERS OF GLOBAL WARMING

Using computers to predict the effects of continued global warming, we can see that increased global temperatures will eventually cause extreme weather all over the world. But extreme weather doesn't just mean extremely hot weather. It means all types of extreme weather: extreme heat, extreme cold, extreme storms, extreme floods, and extreme droughts. How can global warming cause weather outside of just "warming"? Imbalances in Earth's atmosphere can affect the whole system of climate and weather, and that can affect a lot of different facets of life on our planet. Let's take a look at a few examples.

Stronger Hurricanes

One of the measurable effects of global warming is an increase in storms, hurricanes, and tornados. As global warming causes the oceans' water temperatures to rise, the intensity of storms begins to increase. This is because a huge weather system like a hurricane gathers energy from warm waters. This energy is eventually released in the fierce winds of the hurricane. In the

(Opposite page) Hurricane Wilma caused billions of dollars of damage in Florida in 2005. Could global warming make disasters like this all too common?

WEATHER PREDICTION MODELS: HOW WELL DO THEY WORK?

Predicting how global warming will affect the environment is a tricky business. It's not always easy to tell what will happen in the future, especially with something as complicated as Earth's climate. Therefore, scientists use very powerful computers known as global climate models (GCMs) to help them predict the weather.

Those on the myth side of the debate say that data from the GCMs is based on computer climate models that don't accurately represent Earth's climate. They say actual temperature readings like those taken from thermometers, or weather balloons, or even satellites show that in fact our climate is not warming as fast as the global climate models say it will.

Some scientists believe that looking at the actual weather that is taking place now and comparing it to weather from the past is the only way to tell what's actually happening to our climate today. They believe our climate has simply too many factors to take into consideration for even a very powerful computer to predict the future accurately.

Yet, GCM data is still considered by many scientists to be the most reliable predictor that global warming may quickly get out of control. Most of the claims from the crisis side of the global warming argument, including those of the IPCC (Intergovernmental Panel on Climate Change), are based on the findings of weather prediction models such as those taken from GCMs.

1938   1981   2006

As these photos show, Grinnell Glacier at Glacier National Park in Montana has melted considerably since 1938. What will the park be called if all the glaciers melt?

last few years, the Gulf Coast of North America has seen record-breaking hurricanes, both in the number of hurricanes as well as the strength of the hurricanes. And it's not only that hurricanes are getting more intense in places like New Orleans, where they're accustomed to large hurricanes. In fact, hurricanes are beginning to occur in places where they were previously unknown. In 2004, for example, a hurricane hit the country of Brazil for the first time. If things continue to get worse, it's impossible to say how big the hurricanes will get, or where they'll hit next.

The melting of the world's glaciers is also a great problem because it leads to increased sea levels. This means that whole stretches of beach and coastline where people now live could one day be threatened. In fact, if the sea levels get too high, major cities like New York and New Orleans could one day lie completely under water. The rising of sea levels and the melting of glaciers also increase the chance of dangerous tsunamis. Due to melting, massive pieces of glaciers could crack off, sending giant tsunamis toward major cities worldwide. Finally, many rely on glaciers to supply them with fresh water. When these sources disappear, millions will be affected.

Longer, More Intense Droughts and Spreading Disease

The higher global temperatures rise, the greater the droughts, which means it gets harder and harder to grow crops. Global warming could potentially kill off some of the world's supply of rice and other grains, which would mean large-scale famine. Considering the fact that our population is growing every day, it's a very scary development if our food supplies begin to shrink because of global warming.

The more temperatures rise, the more disease we will see worldwide as well. You see, mosquitoes breed faster in warmer climates. The more mosquitoes there are, the quicker diseases like malaria, Ebola, West Nile virus, and other horrible diseases will spread across the world. Some scientists project that two-thirds of the world could be at risk for malaria by 2100 if global warming isn't slowed. Less than half the world is at risk for malaria today.

Global warming could create mass floods, mass droughts, mass storms, and mass destruction. Take all of that into con-sideration and you're looking at a world where entire nations could end up starving, or dying of thirst, or homeless. Who's going to feed them? Who's going to rebuild their houses? Who's going to give them water?

GLOBAL WARMING AFFECTS THE WHOLE EARTH

Global warming doesn't affect only humans; it affects other animals as well. Changes in climate and weather could lead to mass extinction of countless animal species. It took most animals millions of years to adapt to their environments. If global warming quickly transforms these environments, animals may not be able to adapt fast enough to survive. Entire species could go extinct.

We can already see a perilous situation developing with the Arctic polar bear. Polar bears depend on the thick sheets of ice that cover their habitat. They hunt for prey on the ice, they travel to warmer environments on the ice, and they search for fish through holes in the ice. If the Arctic ice melts rapidly, as it is doing now, then more and more polar bears will starve to death or even drown.

Warmer weather will also harm beautiful tropical coral reefs, living environments that are quite sensitive to changes in water temperature. This development could throw off the entire sea ecosystem, potentially killing hundreds of species of fish. In fact, some scientists say that global warming could eventually kill 30 to 40 percent of the world's species. Imagine a zoo with almost half the animal species gone. Imagine a fish tank with only half the species of fish.

If the crisis side of the global warming debate is right, it would be a horrible disaster for our planet—a disaster greater

With large sheets of ice breaking up throughout the Arctic, polar bears are left unable to hunt for food or migrate in their natural patterns.

than any other faced by humans before. Now let's take a moment to catch our breath and hear from the other side of the argument, to find out what others think global warming might do to our planet.

MYTH: GLOBAL WARMING HAS ITS ADVANTAGES

According to those in the myth camp, global warming will have a primarily positive impact on our planet. If you think about it in simple terms, a warmer planet means warmer weather, longer summers, shorter winters, and a milder climate overall. And the milder our climate becomes, the more beach days, water skiing, suntanning, and picnics we can enjoy.

Overall warmer weather also means longer seasons for building new houses and other construction projects, cheaper heating bills in the wintertime, easier travel throughout the year, and fewer winter-related deaths.

Lower Heating Costs

People worldwide could save billions of dollars on heating-fuel bills during the winter. For many who live in very cold environments, the expense of paying for gas, coal, or oil to heat their homes is a great burden. These huge cuts in people's heating bills could provide a boost to the entire world's economy, especially in places where winters take a great economic toll.

Expanded Tourism

Warmer weather also means an increase in tourism worldwide. The longer the summers, the longer the tourism season, which can help bring money to countries whose economies rely on tourism. Warmer weather could also bring tourism to places that never had tourism before. On the flip side, it's important to think about all of the ski resorts and snowy places that rely on tourism to survive as well. Global warming would definitely be hard on winter-related tourism.

Extended Growing Season = Improved Agriculture

A warmer planet could also provide a boost to agriculture worldwide. The warmer seasons are when farmers grow the crops that feed us year-round. If the warmer seasons are longer and the colder seasons are shorter, then there will be more crops grown to feed hungry people around the globe. In addition, if

Warmer weather and shorter winters mean longer growing seasons for farmers. Growing more crops will benefit markets worldwide.

the warming is caused by increased CO_2, this, too, is a good thing. Plants thrive in environments with a good supply of CO_2.

In countries like the United States, where there is less hunger than in impoverished countries, more agriculture means that we will have more food to ship overseas to countries that need it. Also, the more food there is, the cheaper the food becomes. This benefits both the needy countries looking to purchase food to feed their people as well as the moms and dads looking for cheaper produce at the grocery store.

Warmer weather around the globe also means there will be more cultivable land (land suitable for agriculture). Places where once it was impossible to grow crops because it wasn't warm enough could become ideal places for farming. In addition to larger amount of crops, this will also mean more room to plant new trees and forests, which give our planet vital oxygen and help suck up excess carbon dioxide.

Global Warming Creates Opportunities

As we found out in the crisis side of the argument, warmer weather can lead to hurricanes, storms, and tornados that are greater in number and intensity. Humans are good at adapting to a changing environment. So climate change could potentially lead to the invention of new, better ways of predicting and preparing for severe weather. Climate change could push humans to build stronger levees, to build better wind-proof homes, and to build dams that can withstand the stronger hurricanes. The more prepared we are for nature's wrath, the greater our ability will be to prevent another disaster like Hurricane Katrina.

Also, these new storm-related technologies could lead to economic growth. New climate- and weather-related industries might emerge, creating jobs to replace those potentially lost by damaging storms.

The basic view of the myth side of the argument is this: Severe weather is nothing new for our planet. There has always been flooding, there have always been hurricanes, and there have always been extreme hot and cold weather. Just because we're going through a period of intense weather right now doesn't necessarily mean the sky is falling. The human race is strong and has learned to adapt to harsh weather, and it will continue to do so.

Chapter 5

Can Anything Be Done About Global Warming?

Now that you know about the possible effects of global warming on our planet, it's time to take a look at what both sides of the debate think should be done about this issue. Do you believe global warming is a myth that should be ignored? Or is it a crisis that must be solved? Either way, how we come together to deal with this issue will affect the future of us all.

MYTH: NOTHING CAN OR SHOULD BE DONE ABOUT GLOBAL WARMING

If you don't believe global warming is a human-made disaster waiting to happen, why would you do anything

43

to prevent it? This is the belief of most scientists, politicians, and regular citizens who are on the myth side of the argument. In fact, they believe that attempting to prevent global warming could actually cause economic disaster worldwide.

Change Is Expensive

Switching to such newer energy sources as solar energy or hybrid vehicles could be extremely costly. Some new jobs would be created, but millions of workers in developed countries would lose their jobs and their livelihoods. The changes would probably be especially hard on people in less developed countries, where they may not even be able to afford the prices of today's energy sources. Imagine if all the drivers in the world had to give up their cars and buy brand-new cars that used newer forms of energy. Or imagine if solar panels had to be bought to power every house on the planet. It would be more expensive than most people could afford.

What would happen if the entire world was forced to quit using fossil fuels? Some new way of producing energy would have to be invented. We'd have to find a new way to power our cars, heat our homes, and electrify our cities. And as with all new, cutting-edge technologies, this new energy source would most likely be very expensive, at least at first. That would mean only people in the world's wealthy, developed countries could afford the new technologies. Less developed countries could fall even deeper into poverty, which could lead to increased instability, homelessness, and starvation.

What if we made these changes and we were wrong about global warming? What if global warming really wasn't caused

Solar energy can potentially power entire cities without emitting greenhouse gases into the atmosphere. Unfortunately, installing solar panels is costly.

by human activity and wasn't going to become a major problem for our planet? Wouldn't it be tragic to create so much instability and let so many people fall into poverty because of false information? Almost everyone on the myth side of the global warming debate agrees that we need to be sure that global warming is really the fault of us humans before we take drastic measures to repair it.

No Turning Back

Many who believe that global warming is a myth say that even if global warming is caused by humans and is leading us down a dangerous path, there is nothing we Americans could do about it anyway. After all, considering that the whole world

Even if the United States cuts back on CO_2 emissions, developing economies will continue to rely on cheap fossil fuels. In China, energy production from coal is increasing rapidly to meet demand.

relies so heavily on fossil fuels, it would be nearly impossible to reverse global warming.

Stopping global warming would mean that the human race as a whole would have to give up fossil fuels. This would mean no more gasoline for cars and buses, no more coal-generated electricity, and no more fuel for ships or airplanes. Not only that, but once the world gave up fossil fuels, we would also have to find a way to get all of the extra CO_2 out of the atmosphere. How would we do that? Leaving it up to nature could take centuries. If what those in the crisis camp say is true about the effects of global warming, then we're never going to survive long enough to repair our atmosphere anyway.

IT'S A CRISIS: WE CAN STOP GLOBAL WARMING, BUT WE HAVE TO START TODAY

Many who believe global warming is a crisis think that since it took humankind only about 150 years to let global warming get out of control, then surely we can repair our planet in a short time. With specific and immediate policy changes, we can ease the negative impact of the potential disaster.

But here's the bad news: As of right now, 80 percent of the world's energy comes from fossil fuels. If we don't start finding more environmentally friendly energy sources soon, it is projected that 90 percent of the world's energy will come from fossil fuels by the year 2020. Finding a replacement energy source for 80 to 90 percent of the world's energy needs may sound like an impossible task. But according to the many scientists worldwide, it's potentially within our grasp. In fact, energy sources that don't pollute our atmosphere with CO_2 already exist. People pushing for a reduction in greenhouse gases point to the following energy sources as good ways to help reduce (and eventually eliminate) our dependence on fossil fuels.

Solar Energy

The sunlight entering our atmosphere can be transformed into solar energy that can be used to power all forms of technology, including cars. Here's an amazing fact: Every single day, Earth receives enough energy from the sun to power all of our existing technologies for twenty-seven years. Most of that power goes to

THE KYOTO PROTOCOL

The Kyoto Protocol is a treaty drafted by the United Nations in 1997. Its aim is to bring the world together in an effort to reduce greenhouse gases. Countries that signed the agreement pledged to stick to a set of guidelines that would help them reduce their burning of fossil fuels and the production of greenhouse gases. As of this writing, 169 countries have signed the treaty. These countries account for about 55 percent of the world's greenhouse gas emissions.

According to the official government energy statistics, the world's top six CO_2 polluters in 2004 were the United States, China, Russia, Japan, India, and Germany. Russia, Japan, and Germany have signed the

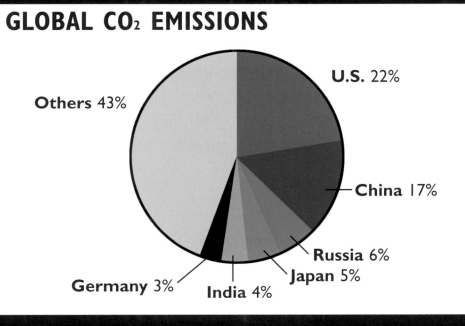

GLOBAL CO₂ EMISSIONS

- **Others** 43%
- **U.S.** 22%
- **China** 17%
- **Russia** 6%
- **Japan** 5%
- **India** 4%
- **Germany** 3%

Kyoto treaty. China and India have also signed the treaty. However, China and India are considered "developing" countries, so they do not have to comply with the treaty's CO_2 reductions. The United States—the world's worst polluter by far—has not signed the treaty, saying it would cripple the U.S. economy.

The Kyoto Protocol would help bring the CO_2 levels in the atmosphere back to where they were in the early 1990s. But even if every country in the world signed on to the Kyoto treaty, it would take a Kyoto treaty nearly twenty times as restrictive to make the amount of CO_2 in the atmosphere not a problem.

waste, but if we figured out better ways to store solar energy, we'd never have to look anywhere else for power.

Hydrogen and Hybrid Vehicles

Our cars play a big part in adding to global warming, pumping out huge amounts of CO_2 into the atmosphere. Hydrogen-powered and hybrid vehicles are two existing technologies that can help us cut way back on our burning of fossil fuels. Hydrogen vehicles burn hydrogen, a gas, in order to power the vehicle. Hybrid vehicles rely partly on regular gasoline and partly on electricity from batteries.

Each windmill in this wind farm in Palm Springs, California, can generate enough energy in a single hour to power a typical American home for a month.

Wind Power

Wind power is already used in limited areas around the world. Wind farms use giant turbines (standing propellers that look a lot like toy pinwheels) to generate power from the wind. Similar to using sunlight as energy, wind turbines produce power without pumping harmful CO_2 into the environment. Unfortunately, current technology is not efficient enough to make wind power a viable alternative to fossil fuels on a large scale.

Other Measures Can Help

Outside of the big changes needed to prevent global warming, there are things individuals can do to reduce the amount of fossil fuels that are burned. Instead of using traditional incandescent lightbulbs, switching to compact fluorescent lightbulbs in homes and workplaces will use much less energy. In addition, individuals can be more conscious of overall energy use. When traveling short distances, walk instead of using a car. Turn the lights off in unoccupied rooms, put on a sweater instead of turning on the heater, and roll down the window in a car or open a window in a room instead of turning on the air-conditioning; all these will conserve energy. Anytime an individual saves some energy, CO_2 is prevented from entering the atmosphere.

BOTH SIDES CAN AGREE ON THIS: THERE IS HOPE

Both sides make valid points in the debate on global warming. However, regardless of whether global warming is caused by humans or not, it is a fact that the world's supplies of fossil fuels—coal, gas, and oil—are limited. So eventually, humans will have to begin looking at alternate fuel resources. For this reason, it is important that people on both sides of the debate come together to conserve energy and begin to make the transition to better, perhaps renewable, sources of energy.

Timeline

900 CE–1100 CE Medieval warming period—Unusually warm period in Earth's climate history, often cited as proof that today's high temperatures are normal.

1800–1870 The Industrial Revolution—Beginning of humanity's heavy reliance on coal and gas to power technologies such as railroads and factories.

1824 Physicist Joseph Fourier sets the groundwork for understanding the greenhouse effect when he states that the world would be a lot colder if it lacked an atmosphere.

1860 The beginning of well-documented global temperatures.

1920–1925 Opening of giant oil fields in Texas and the Persian Gulf, making gas and oil cheaper.

1970 First Earth Day.

1970 Creation of U.S. National Oceanic and Atmospheric Administration, one of the world's most important organizations studying climate change.

1974 A year marked by droughts and severe weather brings climate change to the public's attention.

1974 First signs of damage to our ozone layer caused by CFCs (chlorofluorocarbons).

1974 Media begins to cover global "cooling," thought to be made worse by the use of aerosol spray cans.

1988 Global warming is first discussed by major media outlets.

1990 First report from the Intergovernmental Panel on Climate Change (IPCC) is released; it states that humans are most likely to blame for global warming.

1991 Some scientists state that modern global warming can be blamed on the sun, not human activities.

1995 Second report from the IPCC is published; it contains evidence that global warming is caused by humans and may become a serious problem.

1997 Kyoto Protocol is introduced to the world.

1997 Toyota introduces the Prius, the first mass-produced hybrid car.

2001 Third report from the IPCC is published. Studies show that the amount of CO_2 in our atmosphere is higher than it has been in 400,000 years.

2004 First ever recorded hurricane hits South American country of Brazil.

2005 The Kyoto treaty is signed by more than 100 countries. The United States is not one of them.

2005 Hurricane Katrina hits the Gulf of Mexico and causes mass destruction in New Orleans, Louisiana.

2006 Level of CO_2 in the atmosphere reaches 380 parts per million.

Glossary

aerosols Fine particles or droplets of liquid suspended in the atmosphere. Pressurized aerosol containers are typically used to dispense products in a mist or spray. (Hairspray is an example.)

anthropogenic Word meaning "caused by humans"; often used to describe global warming.

carbon dioxide Natural gas that is an important component of our atmosphere.

deforestation Destruction of large amounts of woodlands and forests.

enhanced greenhouse effect When unnatural amounts of greenhouse gases (mainly carbon dioxide) are emitted into the atmosphere.

fossil fuels Fuels extracted from the earth, such as coal, natural gas, and oil. When burned, fossil fuels release CO_2 into the atmosphere.

glaciers Giant masses of ice, often miles long, that advance and recede in the colder parts of the world.

global climate model Program run on supercomputers that helps to predict weather patterns.

global cooling Theory from the 1970s that claimed aerosols were causing our planet to slowly cool down.

global warming Slow heating of our planet caused by greenhouse gases in the atmosphere, most notably carbon dioxide.

Glossary

ice cores Ancient sections of ice used by scientists to study temperatures and air quality from Earth's distant past.

Industrial Revolution Period in the 1800s in which developed societies shifted toward relying on machinery and mass production to do the labor humans once did.

medieval warming period Era of global warming that took place more than 1,000 years ago.

methane Natural greenhouse gas. Similar to carbon dioxide, it is also found in fossil fuels.

natural greenhouse effect Process by which Earth traps solar radiation, heating the atmosphere.

ozone A natural greenhouse gas.

photosynthesis Process by which plants, trees, and other organisms take in carbon dioxide and give out oxygen.

radically Fundamentally and intensely.

solar energy Technology that transforms sunlight into actual energy that can be used to power a home or a car; usually done using solar panels.

solar radiation Energy emitted by the sun.

sunspots Cooler sections of the sun's surface that, when viewed with a special telescope, appear as black or dark splotches.

tsunami Giant destructive wave caused by a disturbance in the ocean.

For More Information

Environmental Defense
257 Park Avenue South
New York, NY 10010
(212) 505-2100
Web site: http://www.environmentaldefense.org

Greenpeace USA
702 H Street NW
Washington, DC 20001
(202) 462-1177
Web site: http://www.greenpeace.org/usa

Intergovernmental Panel on Climate Change (IPCC)
Geneva, Switzerland
Web site: http://www.ipcc.ch
E-mail: IPCC-Sec@wmo.int

National Environmental Trust (NET)
1200 18th Street NW
Fifth Floor
Washington, DC 20036
(202) 887-8800
Web site: http://www.net.org

National Oceanic and Atmospheric Administration (NOAA)
14th Street & Constitution Avenue NW
Room 6217
Washington, DC 20230
(202) 482-6090
Web site: http://www.noaa.gov

National Resources Defense Council (NRDC)
40 West 20th Street
New York, NY 10011
(212) 727-2700
Web site: http://www.nrdc.org

Sierra Club
85 Second Street, 2nd Floor
San Francisco, CA 94105
(415) 977-5500
Web site: http://www.sierraclub.org

Union of Concerned Scientists (UCS)
2 Brattle Square
Cambridge, MA 02238-9105
(617) 547-5552
Web site: http://www.ucsusa.org

United States Environmental Protection Agency (EPA)
Ariel Rios Building
1200 Pennsylvania Avenue NW
Washington, DC 20460

(202) 272-0167
Web site: http://www.epa.gov

World Meteorological Organization (WMO)
7bis Avenue de la Paix
Case postale No. 2300, CH-1211
Geneva 2, Switzerland
Web site: http://www.wmo.ch

WEB SITES

Due to the changing nature of Internet links, Rosen Publishing
has developed an online list of Web sites related to the subject
of this book. This site is updated regularly. Please use this link
to access the list:

http://www.rosenlinks.com/ad/adgw

For Further Reading

Bailey, Ronald. *Global Warming and Other Eco-Myths: How the Environmental Movement Uses False Science to Scare Us to Death.* New York, NY: Prima Lifestyles, 2002.

Brezina, Corona. *The Industrial Revolution in America: A Primary Source History of America's Transformation into an Industrial Society.* New York, NY: Rosen Publishing, 2004.

The Earthworks Group. *50 Simple Things Kids Can Do to Save the Earth.* Kansas City, MO: Andrews McMeel, 1990.

Ewing, A. Rex, and Doug Pratt. *Got Sun? Go Solar.* Masonville, CO: PixyJack Press, 2005.

Facklam, Margery. *Changes in the Wind: Earth's Shifting Climate.* New York, NY: Harcourt, 1986.

Gershon, David. *Low Carbon Diet: A 30-Day Program to Lose 5000 Pounds.* Woodstock, NY: Empowerment Institute, 2006.

Gore, Al. *An Inconvenient Truth: The Planetary Emergency of Global Warming and What We Can Do About It.* Emmaus, PA: Rodale, 2006.

Green, Kenneth. *Global Warming: Understanding the Debate.* Berkeley Heights, NJ: Enslow Publishers, Inc., 2002.

Hayden, Howard C. *The Solar Fraud: Why Solar Energy Won't Run the World.* Pueblo West, CO: Vales Lake Publishing, 2002.

Johnson, Rebecca L. *The Greenhouse Effect: Life on a Warmer Planet.* Minneapolis, MN: Lerner Publications, 1994.

Langholz, Jeffrey, and Kelly Turner. *You Can Prevent Global Warming (And Save Money!).* Kansas City, MO: Andrews McMeel, 2003.

Parks, Peggy J. *Global Warming* (The Lucent Library of Science and Technology Series). San Diego, CA: Lucent Books, 2003.

Revkin, Andrew C. *The North Pole Was Here: Puzzles and Perils at the Top of the World*. Boston, MA: Kingfisher, 2006.

Silverstein, Alvin. *Global Warming*. Brookfield, CT: Twenty-First Century Books, 2003.

Tanaka, Shelley. *Climate Change* (Groundwork Guides). Berkeley, CA: Groundwood Books, 2006.

Vogel, Carole Garbuny. *The Restless Sea: Human Impact*. New York, NY: Franklin Watts, 2003.

Bibliography

Gelbspan, Ross. *Boiling Point: How Politicians, Big Oil and Coal, Journalists, and Activists Have Fueled the Climate Crisis—And What We Can Do to Avert Disaster*. New York, NY: Basic Books, 2004.

Gore, Al. *An Inconvenient Truth: The Planetary Emergency of Global Warming and What We Can Do About It*. Emmaus, PA: Rodale, 2006.

Green, Kenneth. *Global Warming: Understanding the Debate*. Berkeley Heights, NJ: Enslow Publishers, Inc., 2002.

Haley, James. *Global Warming: Opposing Viewpoints*. San Diego, CA: Greenhaven Press, Inc., 2002.

Maslin, Mark. *Global Warming: A Very Short Introduction*. New York, NY: Oxford University Press, 2004.

McCaffrey, Paul. *The Reference Shelf: Global Climate Change*. Bronx, NY: H. W. Wilson Company, 2006.

Mooney, Chris. "Some Like It Hot." *Mother Jones*, May/June 2005. Retrieved March 9, 2007 (http://www.motherjones.com/news/feature/2005/05/some_like_it_hot.html).

Morgan, Sally. *Science at the Edge: Global Warming*. Chicago, IL: Heinemann Library, 2003.

Index

Index

ABOUT THE AUTHOR

Matthew Robinson is a writer living and working in Los Angeles, California. He is a surfer and bike rider. Robinson drives a hybrid vehicle and does everything within his means to reduce his impact on the environment.

PHOTO CREDITS

Cover (left) National Geographic/Getty Images; cover (right), pp. 9, 25, 33, 45, 46 © Getty Images; p. 5 NOAA; p. 11 NASA Visible Earth http://visibleearth.nasa.gov; pp. 15, 22, 23, 29, 40, 50 © AFP/Getty Images; p. 16 © 2006 MCT/Newscom; p. 18 The Rosen Publishing Group; p. 35 (left) T. J. Hileman/courtesy Glacier National Park Archives, (center) U.S. Geological Survey/photo by Carl Key, (right) U.S. Geological Survey/photo by Karen Holzer; p. 38 © Breck P. Kent/Animals Animals.

Designer: Gene Mollica; **Editor:** Christopher Roberts